I0060135

Understanding Qualcomm® Quick Charge™ 2.0 and the UL Test Program

By Michael Heckrotte

Principle Engineer

UL – Consumer Technology Division

And

George Paparizzos

Staff Product Manager

Qualcomm Inc.

CONTENTS

WHAT IS QUALCOMM QUICK CHARGE 2.0?

Qualcomm Quick Charge 2.0 is a technology used in mobile charging accessories and devices that receive a charge via USB connections.

Building on the innovative Quick Charge 1.0 mobile charging solution, Qualcomm Technologies, Inc. announced Quick Charge 2.0 in 2013 with an expected delivery date beginning in 2014 for devices and accessories. The 2.0 technology, which resides in both the AC/DC wall charger—or other charging device—and the mobile device, delivers a full charge up to 75 percent faster than conventional USB charging. Quick Charge 1.0 is already available in over 100 mobile devices that are based on Qualcomm ® Snapdragon™ processors.

Qualcomm Quick Charge 2.0 Specifications

Quick Charge 2.0 provides significantly higher power over standard USB and micro-USB cables and connectors. Class A technology, which provides 5-volt, 9-volt, and 12-volt inputs in most smartphones, tablets, mobile routers, and other single-cell devices, can provide up to 36 watts over a standard connection. Class B also provides 5-volt, 9-volt, 12-volt inputs, as well as 20V, so it can be used to power Chromebooks, slim notebooks, and any other applications that require higher power.

The minimum current required for charging devices provided with portable devices is 500 mA to insure a decent consumer experience. The Quick Charge 2.0 technology has a current range from 500 mA to 3 amps, depending on connectors. The versatile range lets users power multiple new technologies, some of which have a higher power draw.

The technology is 100% backwards compatible with Battery Charging 1.2, the current standard in the mobile device industry. The Quick Charge devices are backwards and forwards compatible, which means a Quick Charge 2.0 charger can be used with a 1.0 device, or other devices using USB charging. When connected to a 1.0 device, the technology will only allow the voltage appropriate for the 1.0 device. Devices often ship with matching chargers, but 2.0 devices can be charged with conventional USB chargers chargers when necessary. Depending on the implementation, charging a 2.0 device with a conventional charger may increase the time to charge—users will get at most the 1.0 speed. The fastest charge occurs when a 2.0 device is paired with a 2.0 charger.

A detailed specification document exists for Quick Charge 2.0, but is currently only being shared with chipset partners. Chipset partners require the information to generate chipsets with the integrated Quick Charge 2.0 technology. The spec document is not provided to accessory or mobile-device only developers, since the chipsets by themselves support the technology.

Features of Quick Charge 2.0

Quick Charge 2.0 includes all the features users know from the 1.0 technology. The TurboCharge capability maximizes the current provided to the battery via switch-mode operation. Automatic detection of a power source and automatic limiting of input current lets the charger recognize the type of power source, optimize the draw from that source for the device, and creating true universal charging capability.

Charge time is reduced by shortening the taper charging phase, which is done through automatic float voltage control. Low-Rdson FETs improve efficiency, improve thermals, and reduce the number of parasitic losses when charging.

Quick Charge 2.0 also features a High Voltage Dedicated Charging Port, or HVDCP, and automatic input voltage detection that supports backwards-compatible charging for a range of devices.

Quick Charge 1.0 v 2.0

Some key differences between 1.0 and 2.0 include charge current, and typical charge time. Charging time with Quick Charge 1.0 ranges from 1.5 to 3 hours; the 2.0 technology delivers full charge within 45 minutes to 2 hours. The charging time is specific to batter type; a tablet with Quick Charge 2.0 may take more than three hours to charge with certain batteries.

Battery size with Quick Charge 1.0 was 1000 to 3000 mAh, and the charge current from the USB ranged from 0.7 amps to 2.5 amps. Maximum battery size with the Quick Charge 2.0 device is more than 3000 mAh, and the charge current range extends over 2.5 amps.

Supported voltage levels are also different between the two versions, thereby allowing a much higher power transfer for a given current with 2.0.

Improvement in charging time with the Quick Charge 1.0 charger is 40 percent over conventional USB charging. The improvement with the Quick Charge 2.0 charger is as much as 75 percent.

Key Benefits of Quick Charge 2.0

As previously stated, the Quick Charge 2.0 technology allows higher power into the portable device, and as such can deliver up to 75% faster charging versus conventional USB charging. Speed is an enormous factor today for mobile charging devices, because so many individuals rely on their devices for critical computing or communication functionality. A business traveler who runs out of mobile device juice while traveling can use Quick Charge 2.0 charging to bring a Quick Chrge 2.0 enabled device back to life in the time it takes to grab a bite to eat or wait for a flight at the airport. At home or in the office, Quick Charge technology lets users get the most out of mobile devices, relying on them for constant use. Fast charging is even present when the system is operating at full mode, so users don't need to experience downtime to juice up.

Quick Charge 2.0 helps to eliminate the headaches associated with quality of cables and connectors. Higher voltage ranges and flexibility for input voltage ensures performance even if cable length, quality, or thickness are not as expected.

Interoperability is a major plus for Quick Charge 2.0. Consumers don't experience confusion about what power accessory belongs with each mobile device, reducing hassle and the chance of device malfunction due to the wrong power accessory being used. Quick Charge 2.0 is interoperable with existing USB charging implementation, and it works on all mobile devices that accept UBS charge. Users tend to think of tablets and smartphones when using the phrase "mobile device," but Quick Charge 2.0 works with anything that has a Lithium-ion battery, providing true mobile charging capability.

Since Quick Charge 2.0 is part of certain Qualcomm Snapdragon chipsets in production, there are many portable devices on the market that already support the technology. The technology is supported by a wide and fast growing ecosystem. Accessory partners already include makers of wall adapters, car chargers, dongles, docking stations, juice packs, and other items.

Since Quick Charge technology is compatible with USB battery charging 1.2, it complies with the EU mandate for universal chargers.

For More Information about Quick Charge 2.0

Individuals can find information about Quick Charge on the Qualcomm Technologies website. Visit Qualcomm.com for information on chipsets, technologies, and solutions. As additional information is developed, specifically in the area of adaptation of the market, partners will find expanded information and activities on the Qualcomm Technologies website.

QUICK CHARGE 2.0 BENEFITS REQUIRE A TWO-SIDED HANDSHAKE

The interconnectivity and backwards compatibility of Quick Charge 2.0 was previously mentioned. The technology is 100% compatible with existing mobile device and wall adapters, and a two-sided handshake controls each connection and charge interaction.

Consider a standard mobile charging situation as the baseline to understand the importance of the Quick Charge 2.0 handshake. If you connect a regular smartphone or mobile device to a regular USB power accessory, you get 5 volts to around 2 amps in a best case scenario.

Now, consider a situation that involves a one-sided handshake. If you connect a legacy power cord—a regular USB power accessory—to a mobile device with Quick Charge 2.0 technology, you'll get the same input that you got from the old power cord. The old accessory will charge the new device, thanks to interoperability, but you won't get the increase in charge speed. The handshake doesn't occur properly between a Quick Charge 2.0 compatible charger and a 2.0 compatible device—even though the device may request a higher voltage, it will accept what the legacy accessory is able to provide.

The same holds true if you connect a Quick Charge 2.0 power accessory to an old device. The Quick Charge 2.0 power cord is able to deliver a higher voltage and a faster charge time, but the old device isn't capable of receiving. Again, the two-sided handshake doesn't occur. In this case, the Quick Charge 2.0 power accessory defaults to industry-standard voltage.

If both the power accessory as well as the mobile device support Quick Charge 2.0 technology—and there's a lot of momentum with developments in that space—the mobile device can actually benefit from higher power delivery, and therefore, faster battery charging.

It's important to note that the maximum benefits of Quick Charge 2.0 require a two-sided handshake—both the device and the power accessory must include the technology. Still, there are benefits of a Quick Charge 2.0 accessory even for users who don't have many devices with the matching tech. A Quick Charge 2.0 certified power accessory charges a single Quick Charge 2.0 certified mobile device considerably faster than conventional USB charging, and charges most other devices too, reducing the number of accessories required. One of the selling points of Quick Charge 2.0 power accessories is that users can replace the bundle of cords and devices they may have in their desks, drawers, laptop bags, or cars.

The hand-shaking scheme also provides safety benefits and protects equipment, as the technology regulates itself to avoid over delivering voltage. For example, if a traditional wall adapter is used to charge a mobile device that supports Quick Charge 2.0, the portable device receives a traditional 5 volts and whatever current is delivered by the charger. The device is programmed to accept that charge, so there are no issues. Conversely, if the charger is Quick Charge 2.0 enabled and the device is not, the charger isn't going to push 12 or 20 volts on a device that can't handle the load. Instead, the technology stays at a default level, which is 5 volts. That's exactly how traditional power accessories and mobile devices operate today.

Essentially, the technology lets the device and power source communicate over the voltage need. The mobile device asks for particular voltage. If that voltage isn't available—if the charger is Class A and the device is Class B, asking for 20 volts, for example—then the charger doesn't recognize the command for a higher voltage and will remain at the default voltage. At the device end, it could be programmed to ask for the next voltage down for optimal charging with any accessory. This fact is important, because it ensures device safety even when the Qualcomm Quick Charge 2.0 icon (which is covered in the next chapter) is monochromatic or the user is challenged in some way that disallows them to see the color of the icon.

End-Products That Incorporate High Voltage Dedicated Charging Ports

Quick Charge 2.0 technology is incorporated in both the power-supply side and the device side. On the power-supply side, the technology is included in wall or AC adapters, car chargers, juice packs and USB hubs. Almost any accessory that provides power and has a USB port is able to include the technology.

On the device side, the deciding factor is the battery type. Almost any device operating off of a Lithium-ion battery can incorporate the technology, as long as the device is capable of using a USB port for charging purposes. The most common examples in today's market with regard to USB quick-charging technology are going to be mobile devices such as phones and tablets as well as laptops or mobile routers. The technology can also be extended to non-conventional portable applications and even toys.

The technology works with all chipsets. The technology is already embedded within certain Qualcomm Snapdragon chipsets, and Qualcomm offers solutions that can sit next to any other chipset to enable the technology.

Qualcomm Quick Charge 2.0 technology can also be incorporated into wireless charging applications. For example, Qualcomm® WiPower™ technology, a next generation wireless power transfer, is developing transmitter reference designs that can utilize the Quick Charge communication scheme to negotiate with the AC/DC adapter to optimize power delivered to the receiver via the transmitter unit. This allows the end user to utilize the same AC/DC adapter for wired charging and also for wireless charging.

QUALCOMM TECHNOLOGIES'S BRAND AGREEMENT

Qualcomm Technologies selected UL as a certification provider and a process for compliance testing is in place. UL testing provides value to both Qualcomm Technologies and device vendors. Rigorous testing standards protect the integrity and value of Qualcomm Technologies's brand. At the same time, vendors receive the benefit of Qualcomm Technologies marketing and branding, including a logo and icon that distinguish certified fast-charging USB power adapters from standard adapters.

To maintain the integrity, trust, and quality of the Qualcomm Technologies brand, the company opted to make UL the sole-source for testing and certification for Quick Charge 2.0 technology offerings.

Double Path to Certification and Brand Agreement

Before a product can be listed as a certified Quick Charge 2.0 item, complete with the logo and icon, two paths must be completed. The first path is certification through UL testing. Venders must complete and submit the required UL application. UL provides a quote for testing and certification services, which the vendor must accept prior to moving forward. The vendor must provide a sample for review as well as required documentation. UL performs tests and evaluates the device for certification under Qualcomm Technologies's standards. Both the vendor and Qualcomm Technologies receive a report detailing UL's findings.

Outside of the technical certification, a vendor must also contract with Qualcomm Technologies for brand and marketing purposes. Vendors should contact Qualcomm Technologies for brand guidelines and marketing agreements. After agreeing to Qualcomm Technologies's terms and receiving UL certification, the vendor can incorporate Qualcomm Technologies's icons and logos into product documentation and labeling. The icon should also be put on the product itself to show compatibility with the Quick Charge 2.0 standard.

Once both the certification and brand paths are completed, Qualcomm Technologies may list the device on its website as a certified Quick Charge 2.0 product. The logo agreement doesn't expire as long as the partner follows all brand guidelines and trademark agreement content.

Full verses Limited Qualcomm Logo Agreements

Qualcomm Technologies offers two levels of logo use for products, depending on whether items are fully certified or receive verification for internal components or designs. Certification is provided for end products that meet Qualcomm Technologies requirements for Quick Charge 2.0 designations. Certification allows vendors to receive a full logo use agreement for that product from Qualcomm Technologies.

Chipsets, modules, and reference designs may go through a verification process with UL. Qualcomm Technologies administers a limited logo use agreement for items that pass UL verification, allowing vendors to leverage branding for promotional purposes. Verification lets manufacturers of chipsets demonstrate compliance with Qualcomm Quick Charge 2.0 protocol when presenting components to prospective OEMs.

Marketing Agreement Documentation

Product listing and logo agreements require the execution of the Quick Charge 2.0 Trade Agreement and the review and acceptance of the Quick Charge 2.0 brand guidelines. The marketing agreement is not fully effective without successful compliance testing as well as certification or verification from UL.

Benefits of Qualcomm Technologies Logo and Listing

Qualcomm Technologies regularly collaborates with vendors across the globe to manufacture, design, distribute and sell products that are based on industry-leading technology from Qualcomm Technologies, Quick Charge 2.0 technology is one example. Qualcomm Technologies works to provide a strong foundation for the development of ideas and products; the company then works with vendors and partners to provide recognizable branding that enhances sale performance for products, solutions and accessories.

Companies who comply with Qualcomm Technologies marketing and trademark agreements and develop products that pass UL testing benefit from Qualcomm Technologies listings, which provide a potential boost in promotional capability. The listing acts as an introduction to the marketplace and a credential for the product or technology when promoting to OEMs, operators, and consumers. The listing publicly identifies the product as conforming to Quick Charge 2.0 specifications, adding an immediate level of trust.

In addition to the listing, companies are able to use Qualcomm Technologies's graphic icons and logos for further promotional identification. The logo lets consumers easily identify quick-charging devices and accessories, encouraging them to opt for the device that is marked when comparing items from various vendors.

The Qualcomm Technologies Icons and Logos

Each company that signs Qualcomm Technologies marketing agreements and presents a device or component that passes UL testing is able to use the Qualcomm Quick Charge 2.0 icon and logo on devices, product material and promotional documentation.

Both the logo and the icon come in blue or green designs, specifying the class for the device. Class A technology, which provides 5-volt, 9-volt, and 12-volt inputs, receives a green logo. Class B technology, which also provides 20 volts, receives the blue logo. It's important for partners to correctly identify products and devices with the right color logos and icons when applicable to avoid consumer confusion and protect the consistency and trust associated with the Qualcomm Technologies brand.

The icon is a lightning bolt within a circle, providing an easy-to-recognize, simple image that many consumers will associate with charging. The design was developed to be placed on devices next to the USB port, letting consumers know that the port has Quick Charge 2.0 capability. In some cases, it's possible a device, such as a laptop, would have multiple USB ports and not all of them would be equipped with the quick-charge technology. Appropriate icon placement lets the user know which port is best for charging, especially when they have an appropriate matching power accessory. Color-coding and icon placement also helps users quickly establish an optimal charging situation, because they can match devices with the right class of cable or charger.

In some cases, colored icons aren't in line with the partner's own brand or device. Specifically, some chargers are laser marked or etched, which means writing and images on the accessory are monochromatic. Qualcomm Technologies did consider that in their design and consulted with existing partners regarding the two colored icons. Overall, there wasn't an objection to the color in the icon, but manufacturers or vendors that are tied to a monochromatic design may etch the icon without color. The colored logo should still be used on packaging and product documentation to inform the customer at the time of purchase.

Separate from the icon is the logo. The Qualcomm Quick Charge 2.0 logo incorporates the icon image along with the name "Qualcomm Quick Charge 2.0" for optimal branding. Partners can use the logo to distinguish their products on websites. They can also incorporate the logo into packaging, product documentation, and marketing materials as long as all logo use is within trademark and marketing agreement guidelines.

A portable device isn't required to have the Qualcomm Quick Charge 2.0 logo, though the logo is required on references to the technology in any manuals or promotional materials, and it should be included on the package. The icon is essential on the device, however, for identification of compatibility.

WHY UL?

In seeking to create industry-leading technology trends, Qualcomm Technologies collaborates with businesses across the globe. Integrity, trust, and quality are top qualifications for Qualcomm Technologies relationships, and those qualities should be passed on to buyers and end users. To ensure integrity and technical quality of any device bearing the Quick Charge 2.0 icon or logo, Qualcomm Technologies is utilizing UL's experience and resources.

UL has over 100 years of experience working with industries and stakeholders to create safe, efficient, and quality work environments and products. UL tests, inspects, and audits equipment and products from almost every industry, providing well-respected and recognized certifications. As of 2014, over 22 billion UL-certified marks appear on products across the world.

Benefits

UL's relationship with Qualcomm Technologies is not without benefits for vendors and manufacturers. UL delivers fast, efficient, and excellent testing services. Applicants for Qualcomm Technologies's Quick Charge 2.0 certification are provided with a quick, accurate assessment; reports are delivered to both the applicant and Qualcomm Technologies after a maximum weeklong turn-around time.

Like Qualcomm Technologies, UL understands the competitive, fast-paced nature of the mobile device market. Every day of implementation is costly in terms of time-to-market and sales potential, which is why UL works with Qualcomm Technologies and applicants for efficient testing.

UL QUICK CHARGE 2.0 T/C SUBMISSION REQUIREMENTS

Every submission must include a completed Qualcomm Quick Charge UL testing and certification application. Both chipset and end product submissions must include a sample for testing, along with the appropriate payment amount for testing and certification or testing and verification. Each type of test also requires specific documentation.

Chipsets, Reference Designs, and Modules Documentation

For chipsets, reference designs, and modules, documentation requirements consist of a datasheet or similar information that includes identification. The charger must be identified as either Class A or Class B. Class A provides 5V, 9V, and 12V. Class B provides 5V, 9V, 12V and 20V.

Documentation must also identify rated output current at each output voltage. For anything other than a chipset, the IC chip that actually performs the High Voltage Dedicated Charging Port detection must be identified.

End Products

For end products, the documentation requirement is quite similar as for the chipsets. Vendors must identify the class of the charger, the rated output current at each voltage, and the IC chip that performs the High Voltage Dedicated Charging Port detection.

End product documentation should also include a schematic, a Bill Of Materials, and a PC Board layout. Additionally, documentation needs to include a scale drawing of the icon and a photo or drawing showing the location of the icon on the product. The icon is designed to be placed near the UBS port with Quick Charge 2.0 capability.

The Cost of Testing and Certification

A fixed cost of $1,000 applies to all testing processes, regarding of product. Additional listing costs apply to each item type. A fee of $250 for end products is charged for final certification and website listing. The fee for verification and listing for a chipset verification or chipset reference design or module verification is $100.

UL TESTING PROCEDURES

Partners seeking an agreement with Qualcomm Technologies must complete the application for UL testing. At the same time, vendors and manufacturers should interface with Qualcomm Technologies. Certain cycle times are associated with a product launch, including preparing labels, packaging and collateral. Free information exchange isn't generally possible until agreements are signed with Qualcomm Technologies, which can cause problems during the application process and can delay development and launch implementation. Qualcomm Technologies advises that companies interface with them as early as possible in the process.

UL Testing Application Procedure

The application procedure and application is available via a download on the UL website. Potential Qualcomm Quick Charge 2.0 partners should download the application, complete all appropriate sections, and return the application to UL for review. As of spring 2014, the application must be printed out and completed, though UL is working to develop an online form for convenient completion and instant submission.

The application requires information including the business's legal name and address, the name and contact information for the person to receive the certificate, and both technical and non-technical contact information. The form itself only takes a few minutes to complete, and it must be signed by an officer or authorized employee or agent of the business. Once complete, the form can be faxed or scanned in and emailed to UL for review.

Once the application is received, UL will review it and deliver a quote to the business for the testing services. As previously stated, the cost of testing is $1,000 as well as a certification and listing fee of either $100 or $250, depending on whether the business is seeking certification or verification. The quote for services is likely to total $1,100 or $1,250 in most cases. UL aims for a turnaround time of about one business day between submission of an application and receipt of a quote for services.

Testing Provided by UL

Once the quote is accepted by the business and all documentation and samples are received, UL moves forward with testing. As of mid-2014, UL predicts a maximum one-week turnaround time on clean submissions. That means, barring any failures within the testing process, certification can be received within a week or two.

The testing provided by UL on the Quick Charge 2.0 technology is detailed in Qualcomm Technologies's compliance test agreement for the technology. The document is proprietary and is typically only shared with chipset developers. UL will also provide safety and EMC testing, and many of the car chargers and wall chargers will be subject to those requirements as well. Customers can also opt for testing that provides results about energy efficiency, responsible sourcing and other elements.

Since UL wants to make one-stop shopping possible when it comes to technical specification testing, customers can bundle other UL services with the Qualcomm Quick Charge 2.0 specific tests. Bundling would need to be discussed directly with UL.

Qualcomm Technologies Certification is Non-regulatory

Some companies have asked about the impact of Qualcomm Quick Charge 2.0 certification on regulatory requirements. The testing report will not cover needs associated with certain regulatory bodies, as the standard is a Qualcomm Technologies proprietary industry standard. The point of the test is not to prove compliance with regulations, but to ensure devices or components meet Qualcomm Technologies's Quick Charge 2.0 specifications. Because approved devices use the Qualcomm Technologies logo and icon, it's important to ensure a consistent approach across the industry for optimal customer experience.

UL can work with businesses to identify testing needs, including what regulatory requirements may be related to products. Simply because the Qualcomm Technologies testing isn't regulatory in nature doesn't mean that a certain product doesn't require regulatory compliance certification.

Location of UL Testing

As of summer 2014, testing for the Quick Charge 2.0 technology was conducted at UL facilities in Fremont, California, Japan and Taiwan. UL plans to establish additional test labs for the program as the need arises.

No On-Site Inspections

The Quick Charge 2.0 certification or verification doesn't require factory inspections to receive or maintain status. If customers opt to bundle the Qualcomm Technologies testing with UL Safety Certification or responsible sourcing testing, then factory inspection would be as normally required for such bundled services.

Limited Power Source Qualification

It's likely that a power source qualifying for the Qualcomm Technologies testing and certification would also qualify as a limited power source under UL 6950-1. However, UL would look at each individual design before asserting that all requirements were met. There are voltage, current and power limits, as well as some limitations with respect to the current protected devices that UL would have to ensure were maintained.

Retesting Requirements

If a business modifies a model or product currently certified as meeting Qualcomm Technologies specifications, retesting may be required. The decision regarding retesting would depend highly on the modifications made to the design. UL recommends that businesses seeking to maintain certification of a modified design submit the new schematic, bill of materials, and layout.

Highlight the differences from the current model, and UL will work with Qualcomm Technologies to identity retesting needs, if any exist.

Qualcomm Technologies Testing Doesn't Result in UL Listing

The Qualcomm Quick Charge listing is not the same thing as the UL Safety listing. Each listing and test is independent. UL will work with applicants who would like to explore both listing options, and may be able to offer bundle testing and reporting to achieve cost and time efficiency due to cross information. Such arrangements are reviewed on an application-by-application basis so UL can determine what limitations may show up in its information technology system.

FAQS REGARDING QUICK CHARGE 2.0 AND UL TESTING

UL, in conjunction with Qualcomm, hosted webinars and other informational sessions about Quick Charge 2.0 technology, the Qualcomm marketing agreements, and UL's partnership with Qualcomm to deliver testing and certification processes. During the scope of those sessions, partners, vendors, and others in the industry brought up a range of questions, some of which are technical or specific in nature. Those questions, and their answers, are included below for the edification of the reader.

What's the additional Bill of Materials for implementing Quick Charge 2.0 inside a mobile device?

According to George Paparrizos, director of product management at Qualcomm, the additional Bill of Materials on the device side depends on the OEM implementation. On certain implementations, Paparrizos says that the additional Bill of Material will be literally zero. On other implementations, some cost may be seen, which would have to be weighed against benefits by the manufacturer before any decisions are made. In most cases, from Qualcomm's experience in the market, the Bill of Material is, if not zero, negligible.

On the chipset side, some of the Qualcomm chipsets on the market already have the technology. In those cases, the Bill of Material is also zero or close to it.

Is there a minimum wire gauge proposed for the UBS cable?

Mike Heckrotte, the principle engineer with UL's consumer technology division, says wire gauge is typically a design decision made by the OEM manufacturer. The Quick Charge 2.0 specification applies at the USB port, so UL looks at voltage and voltage tolerances at the connector. Typically, a USB Type A connector will be on the product housing itself, whereas an adapter that incorporates an integrated charging cord or integrated USB cord with the Micro B connector would typically require internal voltage designs to meet the specifications at the connector end of the cable.

Does the UL testing support only 5 volt and 9 volt without including 12-volt certification?

One of the main benefits of the Quick Charge 2.0 technology is compliance across all the voltages. That goes back to being able to charge different devices and meet standard needs. Qualcomm wants to avoid issues where a charger isn't compatible with a device or that a device can't accept the charger because voltage is too high. The Class A device will be tested for 5, 9 and 12 volts. The Class B device will be tested for all three voltages plus the 20 volt capability.

Why was 20 volt chosen for the Class B device instead of a more common 24 volt?

According to George Paparrizos, 20 volt is an optimized voltage for laptops. In a survey of the computing market, he notes that many Qualcomm providers deliver devices with voltages between 19.5 and 20.5 volts. There are also wireless charging implementations on the market for future products and releases that are centered around 20 volts.

What is the test procedure for a Quick Charge 2.0 charger with a 12-volt DC power input, and how would that test differ from a test for a charger with both AC and DC input?

According to Mike Heckrotte, the test procedure is the same regardless of AC or DC input. UL will test for responses at the output of the charger. UL discussed dual input scenarios with Qualcomm and it was agreed that testing would be conducted at both the AC input and the 12-volt DC input.

If a wide-range AC power input, such as a range of 90 to 260 volts AC, is presented, must it be tested at multiple voltages?

UL tests at the typical voltage available in the lab.

Does Qualcomm provide a list of all the chipset manufacturers that Quick Charge 2.0 could be compatible with and will the cost be the same for all?

George Paparrizos says that Qualcomm can already provide a list of manufacturers that support the technology. Any product that passes the compliance test becomes part of the list, which is publicly available information. Press releases in the public domain indicate some of the chipset partners as well.

When it comes to cost, the goal of Qualcomm is to have zero cost on the wall adapter for the same power rating or as small of a cost starter as possible. The target is to enable as many partners as possible, but Qualcomm cannot make comments on the details or costs of specific implementations. Each chipset partner would have to address or release such information themselves.

What if limited power systems NOM 12 volt are usually rated at a range of 10.8 volts to 14 volts? How can HVDCP car chargers support 10.8 volts on the lower end?

Typically, that would be handled by the designer of the 12 volt to HVDCP power supply. When making the switch from 12 volt directly to 10.8 volt, 10.8 is the lower tolerance of the 12 output volts of HVDCP. The answer would need to be incorporated into the particular design.

For more information about acoustic pressure testing, or to find out how to submit products for testing, contact UL.

Contact Information

Asia: tel. +852.2276.9150

Europe: tel. +49.694898.10500

North America: tel. +1.847.664.1900

Tel: +1-720-414-5559

Email: consumertechinfo@ul.com

Scan the QR code for more information on the certification process offered exclusively by UL

http://industries.ul.com/blog/qualcomm-quick-charge-2-0/